Electricity

Chris Ollerenshaw and Pat Triggs
Photographs by Peter J Millard

Contents

What's in the toybox?

Have you got any toys that run on batteries? There are quite a few in this toy box. Some came from shops. Some were made at home. Some need only one battery to make them work. Some need more than one.

These things have batteries inside them. How many things in your home need batteries to make them work?

Sometimes when things don't work people say 'Perhaps the battery has run out'. What do you think they mean? Did it look any different before?

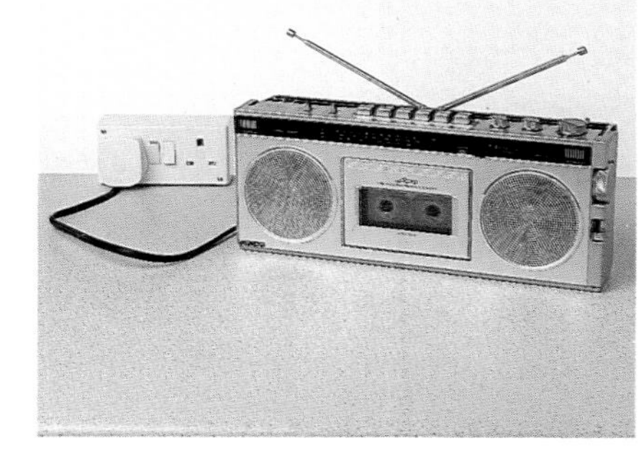

This cassette player has batteries. But it will also work if you plug it in. It can get what it needs to work from its batteries or from the plug. What is it plugging into?

Plugging in

Lots of things that we use every day are plugged in to make them work.

Can you think of some? What about these?

There are so many and we use them so often that we need to be near a supply of what makes them work so we can plug them in easily.

What does make them work? Of course, it's electricity.

Modern houses, schools, offices and factories all use a lot of electricity. They get it from something called the mains supply. How many places in your home or school can you find where you are using the mains supply?

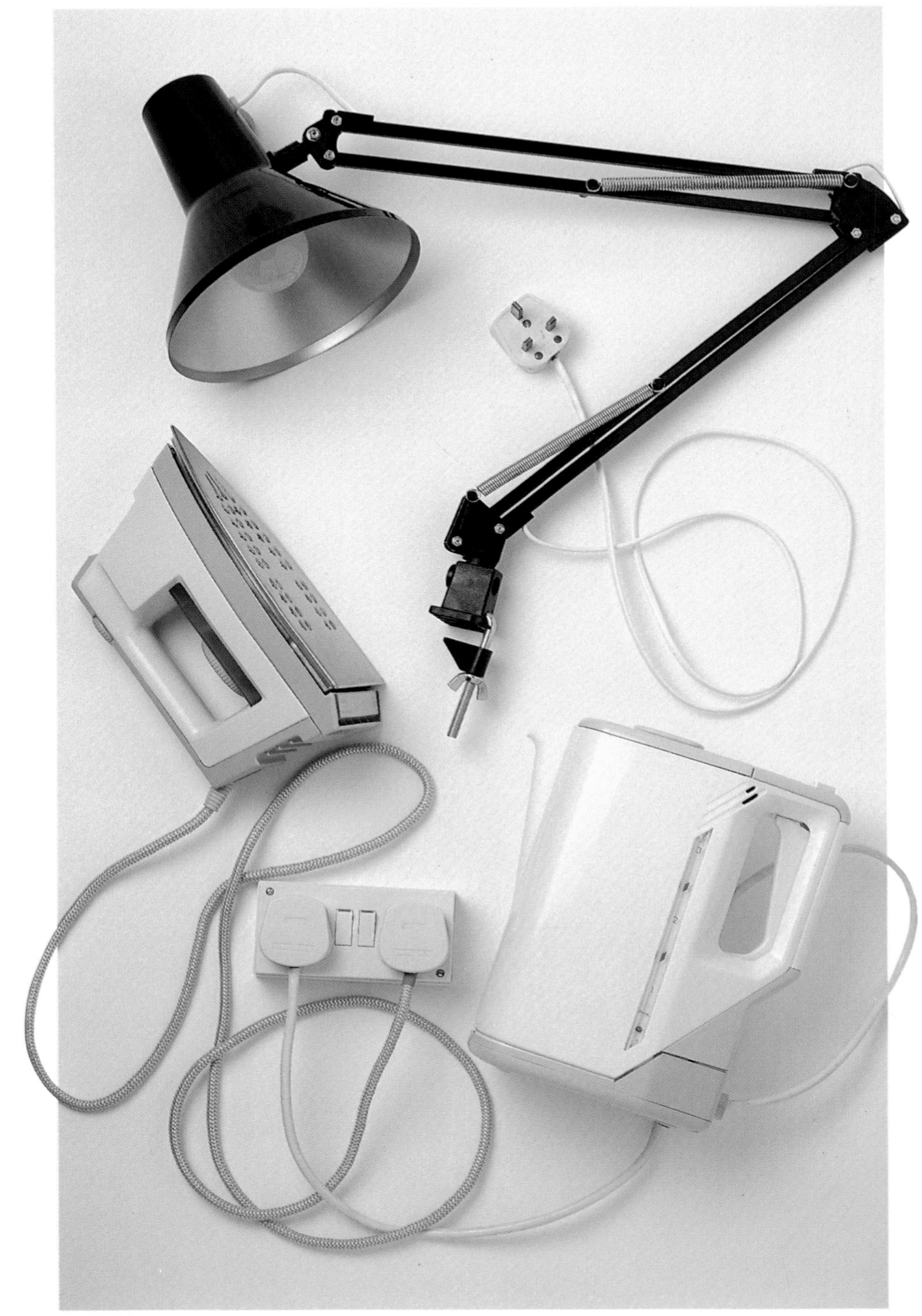

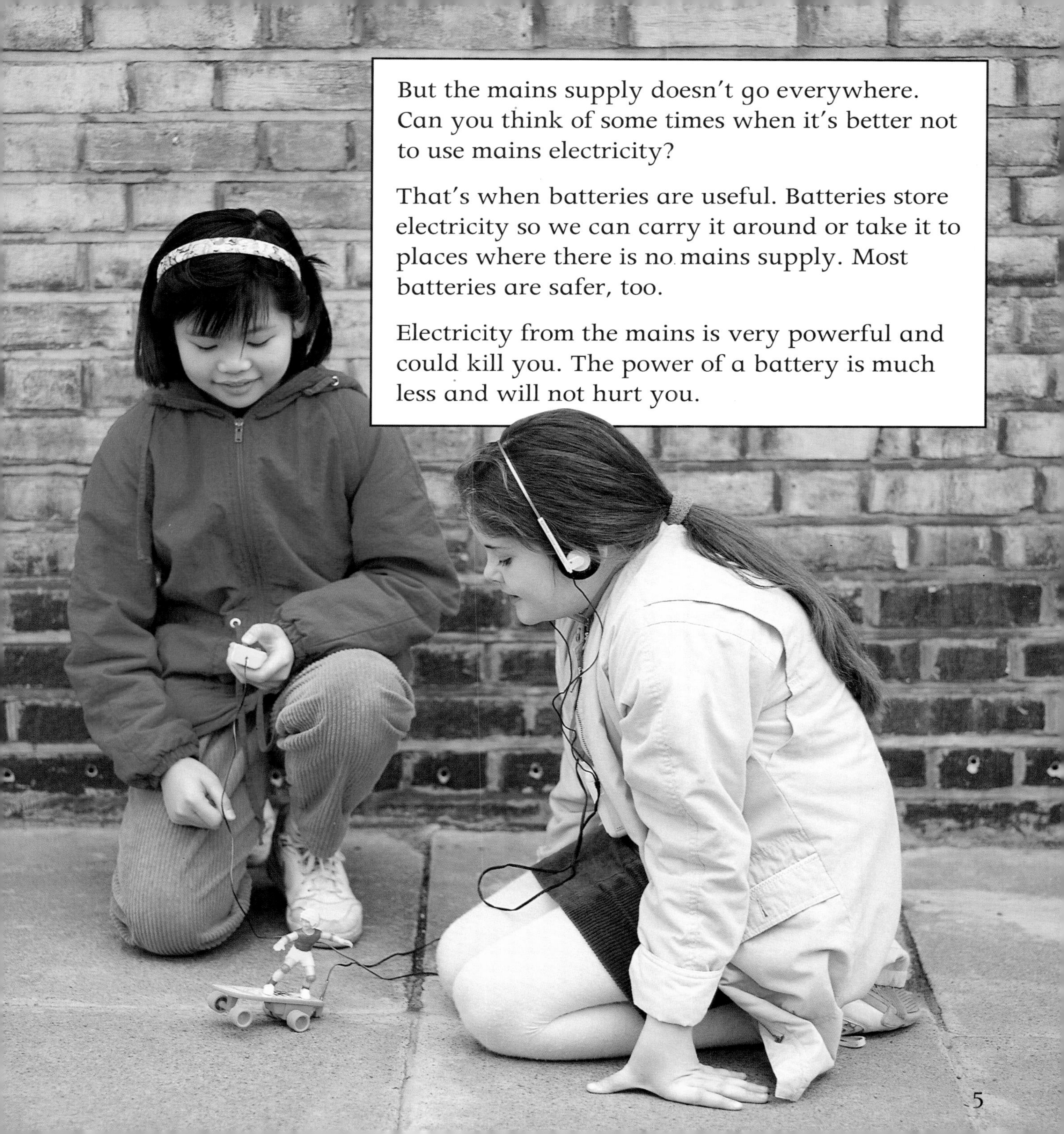

But the mains supply doesn't go everywhere. Can you think of some times when it's better not to use mains electricity?

That's when batteries are useful. Batteries store electricity so we can carry it around or take it to places where there is no mains supply. Most batteries are safer, too.

Electricity from the mains is very powerful and could kill you. The power of a battery is much less and will not hurt you.

What's the problem?

Sometimes toys that use batteries stop working properly. This model lighthouse has stopped working. Its light won't come on. How would you find out why the light isn't working?

You might mend it just by fiddling with different bits. You might make it work by accident. But there's only one way to be really sure that you know what's gone wrong and can work out how to put it right. That is to understand how it works in the first place. You can find out step by step.

If you took the lighthouse apart you could see what it was made of and sort out the electrical parts. Collect a set of the same parts to help you investigate how electricity works.

Start by using just the battery and the bulb. Can you make the bulb light up?

Which part of the battery did you touch with the bulb? Notice *exactly* which parts of the bulb were touching the battery when you made it light up.

Making connections

When you've got the bulb to light up, try again, this time using two pieces of wire. Can you make the bulb light up when it is not directly touching the battery?

Where did you connect the wires to the battery? Where did you connect the wires to the bulb? Draw a picture to show *exactly* where everything touches.

Somehow the electricity is getting from the battery to the bulb. How is that happening?

The battery, bulb and wires are made of different materials. There is paper, metal, plastic, wire and glass. Which of these things is allowing the electricity to flow? Which parts always have to touch? What are they made of?

Use your ideas about what allows electricity to flow to try the next step. This time use the battery, the bulb, the two wires and the bulb holder. (You will need a screwdriver but you won't need to take the screws on the bulb holder right out to make the connections.)

Using the bulb holder

How do you think the bulb fits into the holder? See if you can work out how to connect the wires to the bulb holder. (Hint: remember where you connected the wire and the bulb before.)

Using a part like a bulb holder is very practical. Remember what it was like when you had to make sure all the right parts were touching just by using your hands. The bulb holder has been designed so that it holds the bulb firmly and lets the electricity flow.

When you have connected all the parts and the bulb is alight, draw a picture to show how it is working. Think back to when you were using just the bulb and the battery. Knowing what a light bulb looks like inside will also help you to work out what is going on.

Bridging the gap

You've discovered that electricity flows along the metal wires you have been using. To find out why something electrical isn't working it's useful to know what *doesn't* allow electricity to travel. Is metal the only thing it likes to travel through? Does it matter what kind of metal? What about other materials?

Collect together some different materials and guess which of them you think will allow electricity to flow and which will not let electricity through.

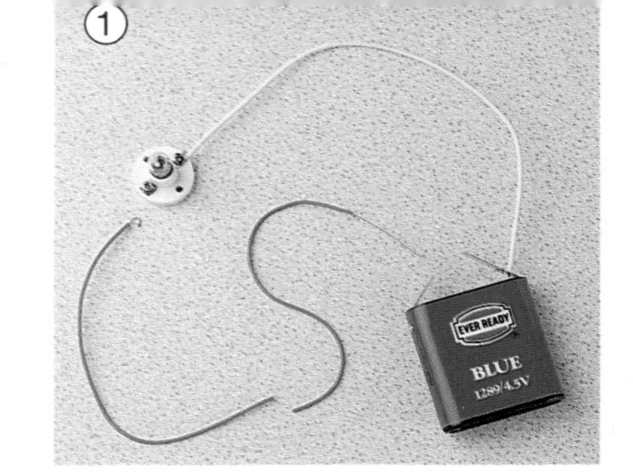

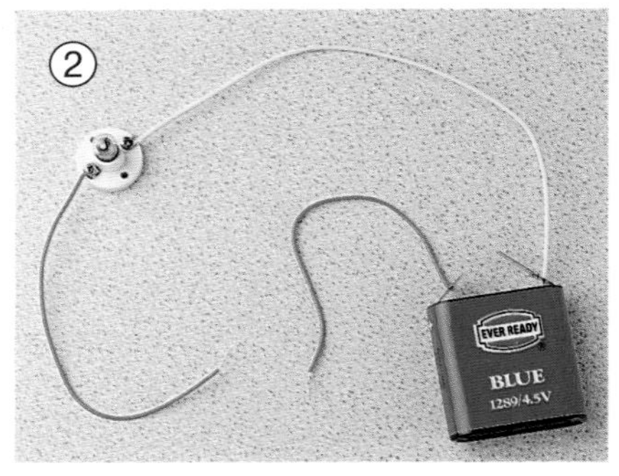

Now disconnect one of the wires from the battery and the bulb holder. Cut it in half (picture 1). (*Never* cut through electrical wires while they are connected.) Strip the wire so that you have enough metal to work with. This is tricky and fiddly. Don't squeeze the scissors too hard! Then reconnect one half of the wire to the battery and the other half to the bulb holder (picture 2).

Try using the different things you have collected to bridge the gap between the wires. Which ones allow the electricity to pass through and light the bulb?
Use a table like this to record what happens.

YES ✓	No ✗
paperclip	rubber band

The materials that allow electricity to pass through are called CONDUCTORS. The materials that stop it flowing are called INSULATORS.

How good were your guesses?

Before electricity

Imagine what life would be like without electricity. Think about how differently people lived before the early 1900s when homes began to have electricity. Look at these household objects which were all used before electricity. Think about people now who live in places where there is no electricity.

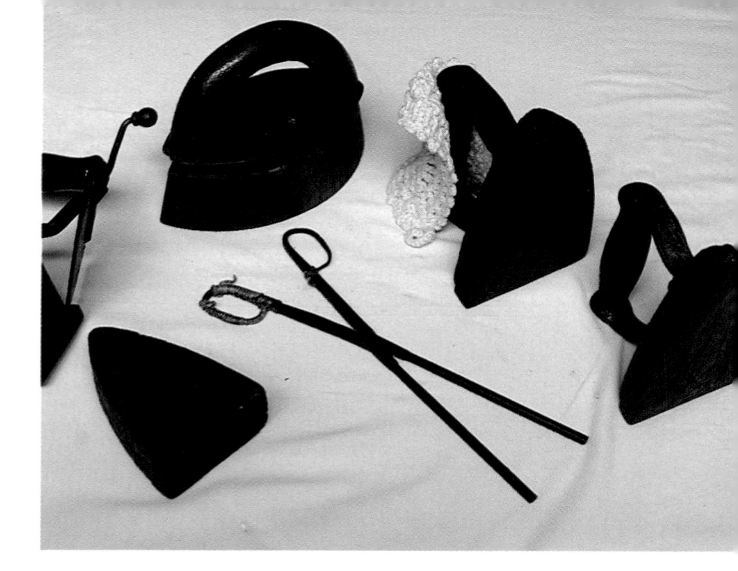

Danger!

Electricity is useful, but although batteries are a safe way of using electricity, mains electricity is very dangerous. Insulators are important because they can be used to make electricity safe. Insulators can wrap up conductors so that people and electricity are kept apart. Which of the materials in your list are used to insulate things like these that we use every day?

People, like metal wires, can be good conductors of electricity. So you must be careful, especially around mains electricity. Make sure you *never* push anything, except a properly insulated plug, into a mains socket.

Safe connections

Have you ever seen someone start a car using 'jump leads'? Sometimes there is not enough power in a car's battery to start the engine so the driver connects it to the battery of another car and uses the electricity from that.

The connections are made with strong crocodile clips (it's not hard to guess why they are called that!) which are made of metal. Can you guess what the handles are made of, and why? (*Never* touch a car battery yourself. They are very powerful and could give you a nasty shock.)

The plugs that we have on every electrical appliance are designed to make good, safe connections with the mains supply. See if you can find an old plug in a junk box at home. Ask an adult to help you to take it apart. Try to find out how it works.

Making a circuit

Now collect all the materials that were good conductors. How many of these conductors can you join together to make a long chain which will allow the bulb to light up? You will need to make sure that everything is touching. Electricity can't cross gaps. It needs an unbroken pathway to flow along.

An unbroken pathway to and from a source of electricity is called a CIRCUIT
A circuit doesn't have to be a circle. Look at the shape of the circuit you've made.

How many things did you manage to include in your chain of conductors? As you added more things was it harder to keep the bulb alight or stop it flickering? The more things you add the more hands you need to hold them in place.

Can you think of a way of making good connections all along your circuit? What materials would you use?

Controlling the flow

It doesn't matter how long your circuit is or what shape it is. If all the connections are good, the bulb will stay alight. But can you see a problem with that arrangement?

As long as the electricity is flowing around the circuit, the bulb will stay alight. To stay alight the bulb is using the electricity made in the battery. When the battery stops moving electricity, the light goes out and the battery is 'dead'. If the bulb is alight when it doesn't need to be, the electricity in the battery is wasted.

The electricity that we use doesn't just happen; it has to be made or GENERATED. Mains electricity is generated in power stations and carried along cables until it reaches our homes. We pay for electricity when we buy batteries or when we pay the bill for the electricity we have used from the mains supply. Electricity is expensive. There has to be a way for us to use it only when we need it.

Could you work out how to stop the flow of electricity? It's all about breaking the connections. So you could disconnect the wires, or loosen the bulb, or pull out the plug. But how practical would that be? Can you think of a better solution?

On-off

Of course. A switch. A switch is something that allows you to control when you want electricity to flow. There are lots of different kinds of switches.

Can you find a way to make a switch?
Use set ups like these to discover what you could use to make an effective switch.

Now you know enough about electricity to find out why the model lighthouse isn't working.

You know the circuit you have just made is working. How would you use these ideas to help you to find out whether the parts of the lighthouse circuit were working or not? You would need to check the battery, the bulb, the connections and the switch.

Looking for clues

You could try the battery first. To do this you would need to take the battery out of your circuit that was working properly and replace it with the lighthouse battery. Don't forget, you would have to check the connections before you switched on.

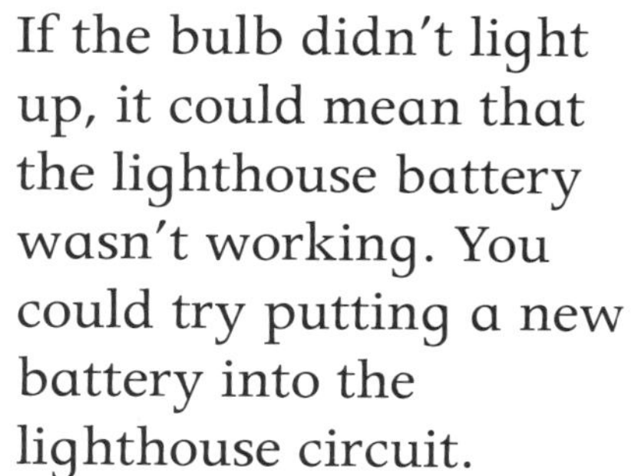

If the bulb didn't light up, it could mean that the lighthouse battery wasn't working. You could try putting a new battery into the lighthouse circuit.

What if the bulb in the lighthouse still didn't light up? You would need to go on checking because sometimes more than one thing can go wrong at the same time.

To find out whether the bulb was the problem, you would need to take the bulb out of your working circuit and replace it with the lighthouse bulb. If the bulb lit up when you switched it on, it could go back into the lighthouse circuit.

Finding the answer

Suppose the battery and the bulb were both working properly but the lighthouse still didn't light up?

You'd need to make sure all the connections were working by checking that the wires were properly connected to the battery and the bulb holder. The bulb would need to be screwed in properly, too.

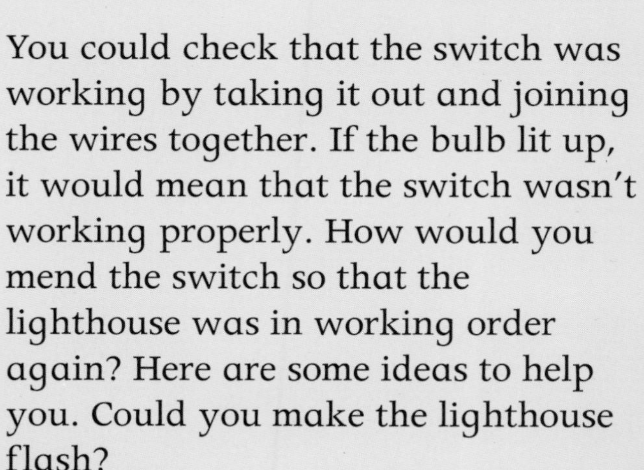

You could check that the switch was working by taking it out and joining the wires together. If the bulb lit up, it would mean that the switch wasn't working properly. How would you mend the switch so that the lighthouse was in working order again? Here are some ideas to help you. Could you make the lighthouse flash?

You could make your own model lighthouse, or a lamp, or light a model room for yourself.

Make a model robot

You can use what you have found out about electricity to build a robot frog. The blueprint for the body is on pages 30 and 31 of this book.

Before you design the circuit that will make the eyes light up you need to investigate what happens when you make two bulbs light up at the same time.

Can you find two different ways of making both bulbs light up together? What difference can you see between these two designs? Can you think why the bulbs in one circuit will be brighter than the ones in the other?

Which circuit design will you choose to fit inside the body you have made from the blueprint? Why do you think that one is better?

Index

This paperback edition published 1999
by A & C Black (Publishers) Limited
35 Bedford Row, London, WC1R 4JH.

First published in hardback 1991.

ISBN 0-7136-5228-4

Illustrations by David Ollerenshaw (blueprint)
and Dennis Tinkler.
Designed by Michael Leaman.

A CIP catalogue record for this book
is available from the British Library.

Filmset by August Filmsetting, Haydock, St Helens.
Printed in Belgium by Proost International Book Production.

Acknowledgments

The photographer, authors and publishers would like to thank the following people whose help and co-operation made this book possible: Michelle, Pui Chi, Nicola, Ebony, Ashley, Mohammed, Larry and the staff and pupils at Avondale Park Primary School, Royal Borough of Kensington and Chelsea.

Robofrog Signaller

TRACE solid and dotted lines onto thin card (a cereal packet will do). Trace the left side of the blueprint first. Next, position one vertical red line on top of the other. Then continue tracing. Take care with dotted lines, especially where they begin and end.

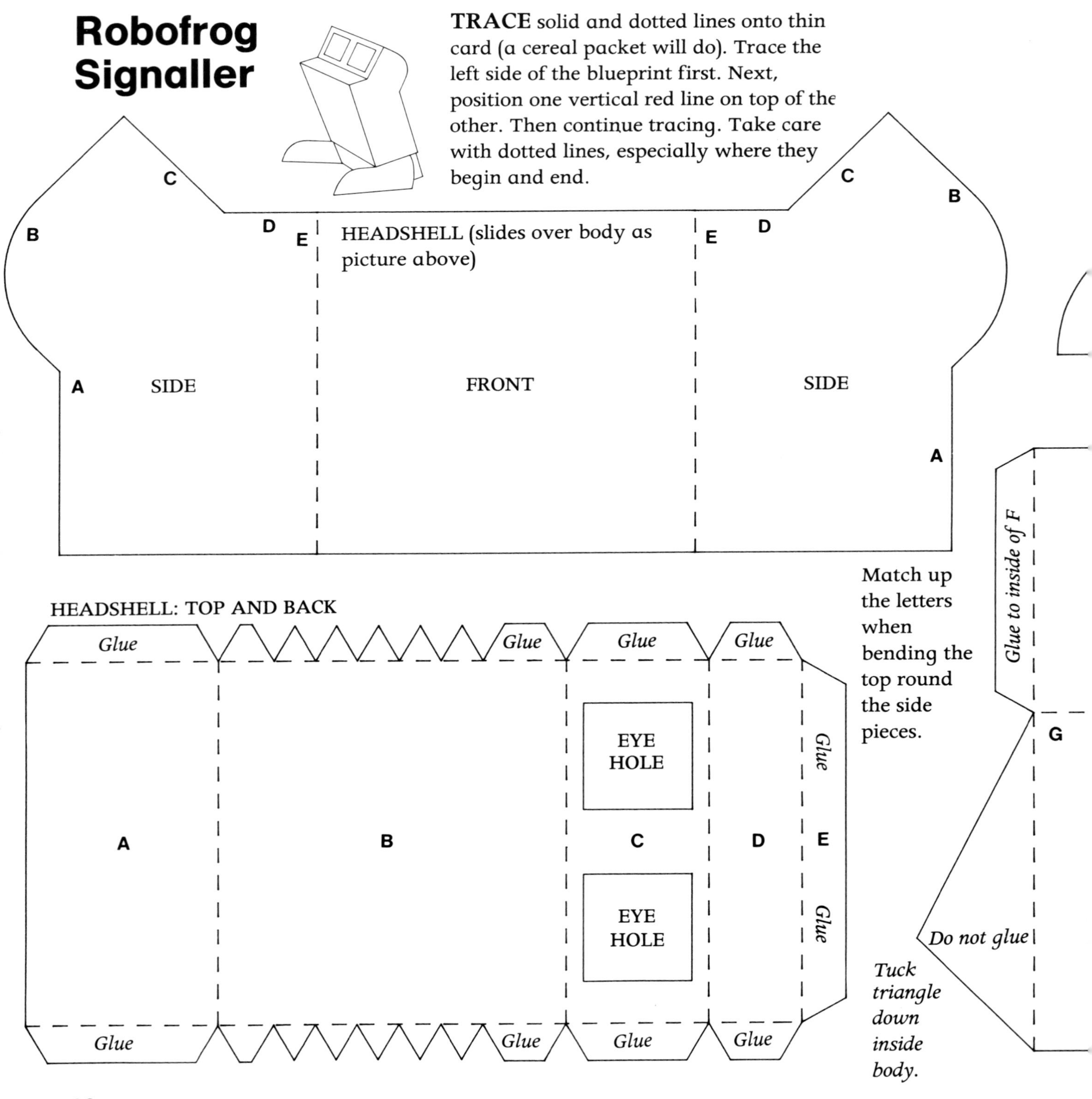

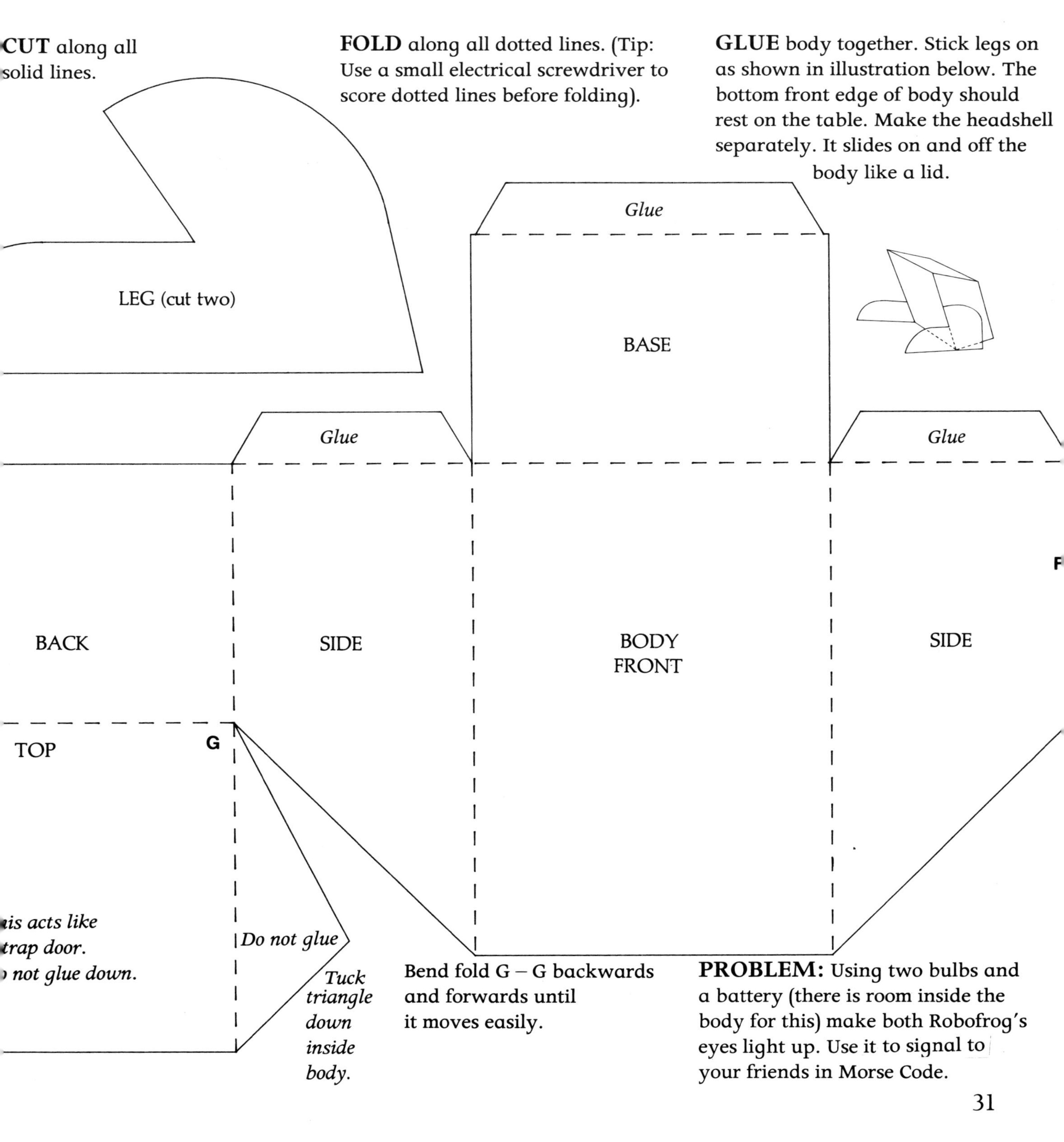
CUT along all
solid lines.
FOLD along all dotted lines. (Tip:
Use a small electrical screwdriver to
score dotted lines before folding).
GLUE body together. Stick legs on
as shown in illustration below. The
bottom front edge of body should
rest on the table. Make the headshell
separately. It slides on and off the
body like a lid.
Glue
LEG (cut two)
BASE
Glue
Glue
F
BACK
SIDE
BODY
FRONT
SIDE
TOP
G
is acts like
trap door.
not glue down.
Do not glue
Tuck
triangle
down
inside
body.
Bend fold G – G backwards
and forwards until
it moves easily.
PROBLEM: Using two bulbs and
a battery (there is room inside the
body for this) make both Robofrog's
eyes light up. Use it to signal to
your friends in Morse Code.

Notes for teachers and parents

Each title in this series promotes investigation as a way of learning about science and being scientific. Children are invited to try things out and think things through for themselves. It's very important for the children to handle the materials mentioned in the books, as only by making their own scientific explorations can they construct an explanation that works for them.

Each Toybox Science book is structured so that it follows a planned cycle of learning. At the **orientation** stage, children draw on their previous experience to organise their ideas. **Exploration** encourages clarification and refining of ideas and leads to **investigation**. At this stage children are testing and comparing, a process which leads to developing, restructuring and replacing ideas. **Reviewing** can occur at the end or throughout as appropriate. Children discuss what they have found out and draw conclusions, perhaps using recorded data. Finally, open-ended problems provide opportunities for **application** of acquired knowledge and skills.

In writing these books we drew on our practical experience of this cycle to select and sequence activities, to frame questions, to make strategic decisions about when to introduce information and specialized vocabulary, when to summarise and suggest recording. The use of real world applications and the introduction of a historical perspective are to encourage the linkage of ideas.

The **blueprint** at the end of each book encourages children to apply their learning in a new situation. There is no right answer to how to get the inside mechanism to work; the problem could be solved in any number of ways and children should be left to find their own.

The national curriculum: the first four books in the series are concerned with energy, forces and the nature of materials explored within an overall notion of movement and how things work.

ELECTRICITY

The 'problem' introduced on page 6 is followed by carefully sequenced activities which will enable children to build up ideas about how electricity works and is used. The book offers a model for the systematic checking of factors, controlling variables and eliminating possible explanations. The fact that electrical energy is brought about by a chemical reaction, used and converted to kinetic energy can be conveyed in terms like "working" and "not working" for batteries (as opposed to "full" and "run out").

The idea of electricity generation by power stations is touched on on page 20. This could be extended by a challenge to find out more from books and other sources. Environmental and ecological issues can also be raised.

Resources

Children working with this book will be best supported by:

- A collection of assorted materials similar to those mentioned in the book.
- A resource box of tools and basics like paper fasteners, rubber bands, etc and everyday junk materials (to be stored and labelled to allow children to access them independently).
- The availability of construction kits.
- Collections of toys and real world objects similar to those mentioned in the book.
- Books and pictures related to the topic of the book to support enquiry and investigation.
- Visits to places where they can see real world applications, current and historical.